Isaac Taylor, William Wrighte, A. Thornthwaite

Grotesque Architecture

Rural Amusement

Isaac Taylor, William Wrighte, A. Thornthwaite

Grotesque Architecture
Rural Amusement

ISBN/EAN: 9783337258283

Printed in Europe, USA, Canada, Australia, Japan

Cover: Foto ©berggeist007 / pixelio.de

More available books at **www.hansebooks.com**

Grotefque Architecture,

OR

RURAL AMUSEMENT;

CONSISTING OF

PLANS, ELEVATIONS, AND SECTIONS,

FOR

Huts, Retreats,	Cafcades,
Summer and Winter Hermi-tages,	Baths,
	Mofques,
Terminaries,	Morefque Pavilions,
Chinefe, Gothic, and Natural Grottos,	Grotefque and Ruftic Seats,
	Green Houfes, &c.

MANY OF WHICH MAY BE EXECUTED WITH

Flints, Irregular Stones, Rude Branches, and Roots of Trees.

The whole containing Twenty-eight new Defigns, with Scales to each.

TO WHICH IS ADDED,

An Explanation, with the Method of executing them.

BY WILLIAM WRIGHTE, ARCHITECT.

A NEW EDITION.

LONDON:

Printed for I. and J. TAYLOR, at the Architectural Library, nearly oppofite Great Turnftile, Holborn.

M.DCC.XC.

PLATE I.

PLAN and elevation of a hut, to be built with trunks of trees and irregular timber. The inside walls may be lined with mofs, and covered on the top with thatch. It is intended to reprefent the primitive ftate of the Dorick Order, and is proper to be placed at the entrance of a wood, or on the top of a fmall eminence. The dimenfions are figured on the plan.

PLATE II.

Plan and elevation of an hermetic retreat, to be compofed of roots and irregular branches of trees, cemented together with a ftrong binding clay, and may be thatched or covered with branches of trees twined round with ivy. The dimenfions are figured on the plan.

PLATE III.

Elevation of an hermit's cell, with ruftic feats attached, eight feet fquare in the infide, which fhould be fituated in a rifing wood near fome running water, to be built partly of large ftones and trunks of trees, fet round with ivy, and lined with rufhes, &c. The roof fhould be covered with thatch, and the floor paved with fmall pebble ftones or cockle fhells. The feats attached are intended to be compofed of large irregular ftones, roots of trees, &c.

PLATE IV.

Plan and elevation for an hermitage in the eaftern ftyle, fuppofed to be built round a tree which fupports its roof; over the door is a ta-blet, with an Arabic infcription; the roof is covered with thatch, in the Chinefe tafte; the infide to be lined with billet wood and mofs. It is lighted from the lanterns above. A. fhould be a couch; B. C. are feats of retirement. The dimenfions are figured in the plan. The ruftic feats on the fide are intended to be compofed of large rough ftones and roots of pollard trees cemented together.

PLATE V.

Plan and elevation of a winter hermitage, in-tended as a retirement from hunting, fowling,

or any other winter amufement; the walls to
be built of flints or rough ftones, and lined with
wool or other warm fubftance intermixed with
mofs, and fhould be fituated on a rifing ground
planted with evergreens.

PLATE VI.

Plan and elevation of a fummer hermitage,
defigned to be in a wildernefs or thick wood ;
the walls to be compofed of large ftones, and the
ends faced with flints ; the roof covered with
thatch, and an owl carved on the top ; the floor
fhould be paved with fheeps marrow-bones placed
upright, or any other pretty device intermixed
with them.   A. is for a couch ; B. C. are feats
of retirement.

PLATE VII.

Plan and elevation of an hermitage in the
Auguftine ftyle ; the front is ornamented with
a portico of palm trees, in the pediment is a
fcull, and a tablet with an infcription.   A.
A. are paffages of evergreens leading to the
two circular retreats, one of which is intended
as a library, and the other a bath ; the tops of
them are intended to be thatched ; b b b. are
niches for feats cut in the evergreens.   This
defign is calculated to be built on a fmall ver-

dant

dant amphitheatre, near a murmuring ſtream, and as a proper retreat from the fatigues of a ſultry day.

PLATE VIII.

A plan, half an elevation, and half a ſection, of a rural circular hermitage, deſigned for an open ſituation near ſome rivulet, planted with weeping willows, &c. The inſide is lighted by a gazebo, ſupported by eight trunks of trees twined about with ivy. · The dimenſions are figured on the plan.

PLATE IX.

Plan, elevation, and ſection, of a grotto in a modern architectonic ſtyle, ornamented with jet d'eaux, ſea weeds, looking-glaſs, fountains, and other groteſque decorations. The dimenſions may be known by the ſcale and the figures on the plan.

PLATE X.

Plan and elevation of a Gothic grotto with four cloſets five feet ſquare; the outſide to be compoſed of flints and irregular ſtones, and ſtudded with ſmall pebbles; the inſide to be ornamented with ſhells, ores, &c. and if built upon an eminence, it would have a very pleaſing appearance.

PLATE

P L A T E XI.

Plan and elevation of an open Chinefe grotto, to be placed at the head of a grand canal, with a bath (A), and a Chinefe temple (B), attached; the arcades to be ice or frofted work; the outfide of the bath and temple to be ornamented with beautiful fhells in the Mofaic tafte; the infide to be groined over, as on the plan, and ornamented with fhell-work and other beautiful incruftations. The whole extent is 75 feet.

P L A T E XII.

Plan and meafures to plate xiii. and xiv.

P L A T E XIII.

Elevation of a Gothic grotto, with cafcades and wings attached (*fee the plan, plate* xii.). The entrance is a faloon of 20 feet fquare; the angles are couped with off niches, where grotefque ftatues or vafes fhould be placed. It is intended to have a fountain in the centre, with antique figures fpouting out water; the walls fhould be lined with flints, decorated with ice-work; the whole is lighted from the gazebo on the top. A. B. are the plans of the two wings or repofitories, which are each defcended to by a flight of four fteps. A. is intended to be ornamented with curious fhells, gems, coral, &c. with fta-

A 4   tues

tues in the niches. B. is to be groined over in
the Gothic manner, with a pier in the centre,
to be lined with flints, intermixed with shells,
looking-glass, &c. The groins should be in-
crusted with frosted work, in the manner of
dropping icicles. Both these wings are lighted
from the rose arches, as appears in the elevation;
the outside to be composed of rough stones in-
crusted and studded with pebbles, shells, &c.
There are placed in the recesses Gothic figures.
The situation should be in some retired copse,
shaded by an adjacent hill, near some murmuring rivulet, where the cascades, or rather fountains, as in the design, may be easily effected,
The measures are marked on the plan.

P L A T E XIV.

Elevation of a rural grotto (*see the plan, plate*
xii.), which should be built of large rough
stones rudely put together, so that the building
may as near as possible imitate the beautiful appearance of nature. If the dome was to be
richly ornamented with pendentive shell and
frosted work, it would look very elegant. In
the middle niche is Neptune on a rock, pouring
out water, which descends under the pavement
through an arch, and forms a running stream.
The side niches are ornamented with satyrs and
other grotesque figures. The situation should
be in a morass, near some water.

P L A T E

P L A T E XV.

A defign for a cafcade or cataract of a great fall of water, decorated with rock-work, fea lions pouring out fountains of water ; and a triton, by way of embellifhment, in the centre.

P L A T E XVI.

A defign for a triumphal cafcade of four falls of water. If care is taken to erect this arch with rude and irregular flints, &c. at the fame time paying a due obfervance to nature, it will have a very magnificent appearance, and look extremely elegant ; and would be a fuperb ornament in a nobleman's park where there is a great fupply of water.

P L A T E XVII.

A grotto, canal, and cafcade, decorated with rock-work, tritons, fibyls, &c. pouring forth fountains of water. The author hopes he may be indulged with obferving, that he hath with great pleafure feen a fine piece of water in the park of the Earl of *Effex*, at *Caffiobury*, near *Watford, Herts* ; and flatters himfelf that if the arch in this defign, on which the triton is placed, was to be executed there in the nature of a bridge, it would have a very magnificent and pleafing appearance.

P L A T E

P L A T E XVIII.

A romantic bridge, or a cafcade of three fheets of water, defcending through arches of artificial rock-work, incrufted with fhells, corals, fea-weed, mofs, &c. and two fea gods lying on their oozy couch, pouring out water.

P L A T E XIX.

Plan and elevation of a ruftic feat for a garden or park, intended to terminate a view. It would look very pretty if it was built with flints, or irregular rude branches and roots of trees.

P L A T E XX.

Plan and elevation for a grotefque or rural bath, very proper to be built in gardens, &c. for the benefit of bathing. It is intended to have three feats within, by way of clofets, for the conveniency of dreffing and undreffing. If the water in the plan be left out, it will look very pleafing as a rural hut.

P L A T E XXI.

Plan and elevation of a rural mofque with minarets. It is divided into an octagon faloon, fupported by eight columns, lighted from the dome. The other apartments are four regular fmall rooms or clofets, which will ferve for various

rious purpoſes. The minarets are placed in the
plan by way of ornament, to ſhew the true taſte
of the Turkiſh buildings; and the ſingularity of
the ſtyle of architecture is ſuch, that will render
it a very pleaſing ornament, if executed in a
pleaſure ground, or upon an elevated verdant
amphitheatre. It may be built of wood, and
ſtuccoed; the inſide ſhould be painted with various rich colours, which would have a pleaſing and elegant appearance. The dome is ſupported by irregular branches of trees, well con-
nected and cramped together. The minarets
ſhould be ſolid, and the pedeſtals (A.B.) ſhould
be decorated with Arabic inſcriptions. For a
more intelligible and hiſtorical account of theſe
buildings, I muſt refer the reader to Dr. *Shaw*'s
Account of *Barbary*, *Le Brun* and *Tournefort's*
Voyage to the *Levant*, &c.

PLATE XXII.

Plan and elevation of a circular moſque
twenty feet diameter, with four cabinets attached, eight feet ſquare; two of which may
ſerve for entrances, having each a ſmall fountain,
five feet diameter; the other two may be for the
purpoſes of ſtudy or uſe. The four minarets at
the angles bring the plan upon a ſquare of forty
feet. The cabinets, as well as the moſque, are
crowned with domes, which ſhould be gilt on
the

the outfide. The great dome is fupported by eight columns, over which are groined arches; an iron baluftrade runs round the outfide, which may be painted blue, and gilt; on the top of the great dome is a light cupola, fupported by eight fmall columns, from whence hangs a chandelier to light the infide when required. The other decorations may be feen in the fection, plate xxiv.

PLATE XXIII.

Plan and elevation for another mofque, with two minarets attached to the body of the building, which may be executed in brick of 14 inches thicknefs. The front is a portico of four columns, in the oriental ftyle, in the centre of which is a fountain for fabateons; which may be feen in the fection, plate xxiv. The niches in the front fhould have Arabic infcriptions in gold letters. The portico is covered with three little domes, in the Turkifh manner, ornamented with crefcents, &c. The infide is lighted from the circular windows and little arches above, which fupport the dome. For the interior decorations fee the fection, plate xxiv. It would look very beautiful if built on an open lawn, planted round with a few cyprefs or other exotic trees. The dimenfions are figured on the plan.

PLATE

PLATE XXIV.

Sections and fcales to the plates xxii. and xxiii.

PLATE XXV.

Two plans of morefque temples to plates xxvi. and xxvii. with their proper meafures.

PLATE XXVI.

Elevation of a beautiful morefque temple (*fee the plan, plate* xxv.). The coupled columns fupport an arcade of interfecting femi-ellipfes, which goes quite round the temple. In the fpandrells are Moors heads, with crefcents, rofes, and ftars, over which is a parapet baluftrade of net or lattice-work. The body of the temple is twenty feet diameter, crowned with an open lantern, from whence it is lighted; the outfide of which is adorned with ftars of glafs on an azure ground. On the top is a pine, which fhould be double gilt; and if the outfide was covered with a glofly fubftance, it would have a very pleafing and magnificent appearance. The ftyle of architecture is a medium between the Chinefe and Gothic, having neither the levity of the former nor the gravity of the latter. The particularities of both this and the following defign are taken from thofe famous remains of Barbarian antiquity, the palace of *L'Lhambra*, at *Granada*, the ancient morefque mofque

at

at *Cordova*, the old caffavee or palace of the
Moorish kings at *Mæquanez*; for the accounts
of which the reader is referred to *Willughbuy's*
Travels into *Spain*, *Ocley's* Account of South
or Weft *Barbary*, and *Shaw's* Travels to the
Levant.

PLATE XXVII.

Elevation of a morefque pavilion *(fee the plan,
plate* xxv.) in the ftyle of the ancient *Moors*,
raifed on three fteps. Over the arches are *Moors*
heads and feftoons. In the middle is a circular
or geometrical ftair-cafe, leading to the top, or
baluftrade. It is crowned with a fquare cupola,
mounted with a morefque ftandard; and is very
proper to be built on an eminence to command
an extenfive view.

PLATE XXVIII.

Plan and elevation for a green-houfe of the
grotefque kind, faced with flints and irregular
ftones. The dimenfions may be found by the
fcale.

Books on *Architecture, &c.*

Printed for I. and J. TAYLOR, at the Architectural
Library, No. 56, High Holborn.

1. *THE Rudiments of Ancient Architecture*; in two Parts: containing an Historical Account of the Five Orders, with their Pro-
portion, and Examples of each from the Antiques : Also, *Vitruvius*
on the Temples and Intercolumniations, &c. of the Ancients ; calculated for the Use of those who wish to attain a summary Knowledge
of the Science of Architecture ; with a Dictionary of Terms : illuf-
trated with ten Plates, and a Portrait of the celebrated *James Stuart*,
Efq. Price, in boards, 5s.

2. Plans, Elevations, and Sections of Buildings, executed in the
Counties of *Norfolk, Suffolk, Yorkshire, Wiltshire, Warwickshire, Staf-
fordshire, Somersetshire, &c.* By *John Soane*, Architect, Member of
the Royal Academies of Parma and Florence. Dedicated, with per-
miffion, to the King. On Forty-feven folio Plates. Price, on Royal
Paper, 2l. 2s. on Imperial Paper, 2l. 12s. 6d.

3. Plans, Elevations, and Sections, of the *House of Correction* for
the County of *Middlesex*, to be erected in Cold-Bath Fields, London ; together with the Particular of the feveral Materials to be con-
tracted for, and manner of ufing the fame in building.

N. B. T is Work is engraved from the original Defigns, and pub-
lifhed with the authority of the Magiftrates, by *Charles Middleton*,
Architect, engraved on 53 plates, imperial folio, price 2l. 12s. 6d.
half bound.

4. *The Cabinet-Maker and Upholfterer's Guide* ; or Repofitory of Defigns for every article of houfehold furniture, in the neweft and moft
approved tafte. The whole exhibiting near three hundred different
defigns, engraved on one hundred and twenty-fix folio plates : from
drawings by *A. Hepplewhite & Co.* Cabinet-Makers, 2l. 2s. bound.

5. *The Builder's Price-Book* ; *containing a correct lift of the prices al-
lowed by the moft eminent furveyors in London to the feveral artificers con-
cerned in building* ; *including the journeymen's prices.* A new edition,
corrected, with great additions, by an experienced furveyor, 2s. 6d.
fewed.

6. *Familiar Architecture* ; confifting of original Defigns of Houfe,
for Gentlemen and Tradefmen, Parfonages and Summer Re'reats ;
with Back-Fronts, Sections, &c. together with Banqueting-Rooms
and Churches. To which is added, the Mafonry of the Semicircular,
and Elliptical Arches, with practical Remarks. By the late *Thomas
Rawlins*, Architect. On fifty-one Plates, Royal Quarto. Price 1l. 1s.

7. *Crunden's Convenient and Ornamental Architecture* ; confifting of
original Defigns for Plans, Elevations, and Sections, beginning with
the Farm-houfe, and regularly afcending to the moft grand and magnificent Villa ; calculated both for town and country, and to fuit all
perfons in every ftation of life ; with a Reference, and Explanation
in Letter-prefs, of the ufe of every room in each feparate building
and the dimenfions accurately figured on the Plans, with exact fcales
for the meafurement ; elegantly engraved, on feventy Copper-
plates, 16s. bound.

8. *The Country Gentleman's Architect*, in a great variety of new Designs for Cottages, Farm-houses, Country-houses, Villas, Lodges for Park or Garden Entrances, and ornamental Wooden Gates; with Plans of the Offices belonging to each Design ; distributed with a strict attention to convenience, elegance, and economy. Engraved on thirty-two Quarto Plates, from Designs drawn by *J. Miller*, Architect, 10s. 6d. sewed.

9. *Garret's Designs and Estimates for Farm-houses*, for the Counties of York, Northumberland, Cumberland, Westmoreland, and the Bishoprick of Durham. Folio, 5s. sewed.

10. *Dr. Brook Taylor's Method of Perspective made easy, both in Theory and Practice*; in two Books; being an attempt to make the Art of Perspective easy and familiar, to adapt it entirely to the Arts of Design, and to make an entertaining Study to any Gentleman who shall choose so polite an Amusement. By *Joshua Kirby*, Designer in Perspective to his Majesty, and Fellow of the Royal and Antiquarian Societies. Illustrated with thirty-five Copper Plates, correctly engraved under the Author's inspection. The third Edition, with several Additions and Improvements. Elegantly printed on Imperial Paper, 1l. 10s. half bound.

11. The same Work in two Volumes Quarto, 1l. 1s.

12. *The Perspective of Architecture*, a work entirely new : deduced from the principles of Dr. Brook Taylor, and performed by two rules of universal application : illustrated with seventy-three plates. Begun by command of his present Majesty when Prince of Wales. By *Joshua Kirby*, Designer in Perspective to his Majesty, and Fellow of the Royal and Antiquarian Societies. Elegantly printed on Imperial Paper, 1l. 16s. half bound.

13. *The Description and Use of a new Instrument called the Architectonic Sector*, by which any part of Architecture may be drawn with facility and exactness. By *Joshua Kirby*, Designer in Perspective to his Majesty, and Fellow of the Royal and Antiquarian Societies. Illustrated with twenty-five Plates. Elegantly printed on Imperial Paper, 1l. 1s. half bound.

14. The two Frontispieces, by Hogarth, to Kirby's Perspective, may be had separate, at 5s. each.

15. *Designs in Architecture*; consisting of Plans, Elevations, and Sections, for Temples, Baths, Cassines, Pavilions, Garden Seats, Obelisks, and other Buildings : for decorating Pleasure-grounds, Parks, Forests, &c. &c. By *John Soane*. Engraved on thirty-eight Copper-Plates, Imperial Octavo, 6s. sewed.

16. *The Temple Builder's most useful Companion*; containing original Designs in the Greek, Roman, and Gothick taste. By C. *T. Overton*. Engraved on fifty Copper-plates, Octavo, 7s. sewed.

17. *The Carpenter's Treasure*; a collection of Designs for Temples, with their Plans, Gates, Doors, Rails, and Bridges, in the Gothic taste, with Centres at large for striking Gothic Curves and Mouldings, and some Specimens of Rails in the Chinese taste, forming a complete system for rural decorations. By *N. Wallis*, Architect. Engraved on sixteen Plates, Octavo, 2s. 6d. sewed.

Plate 1.

A Primitive Hut.

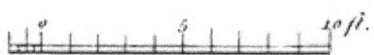

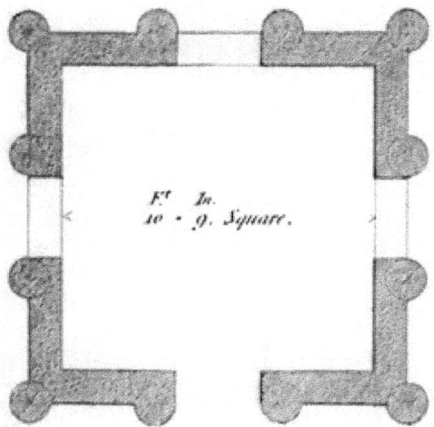

Ft. In.
10 · 9. Square.

Pl. 2.

Hermetic Retreat.

4 f.ᵗ

10 ft. Square.

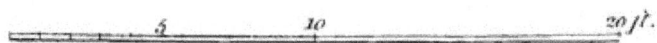

5   10   20 ft.

Pl. 3.

Hermits Cell.

Pl. 4.

Oriental Hermitage.

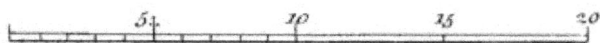

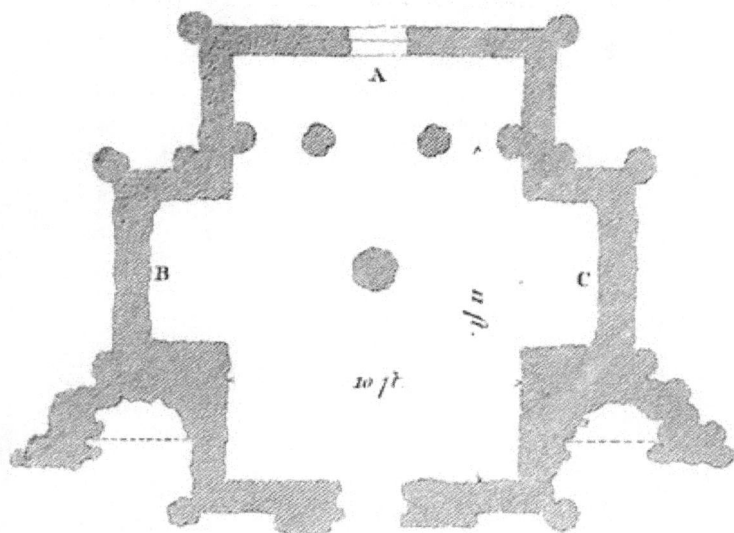

Winter → *Hermitage*   *Pl. 5.*

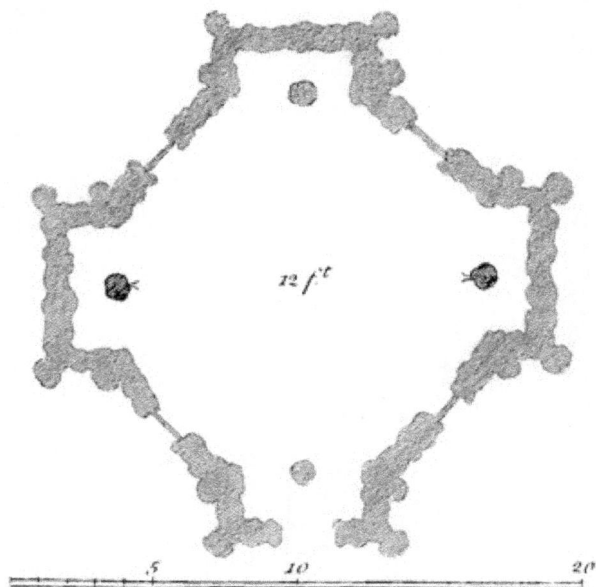

12 f.t

5   *10*   *20*

Pl. I.

Summer Hermitage.

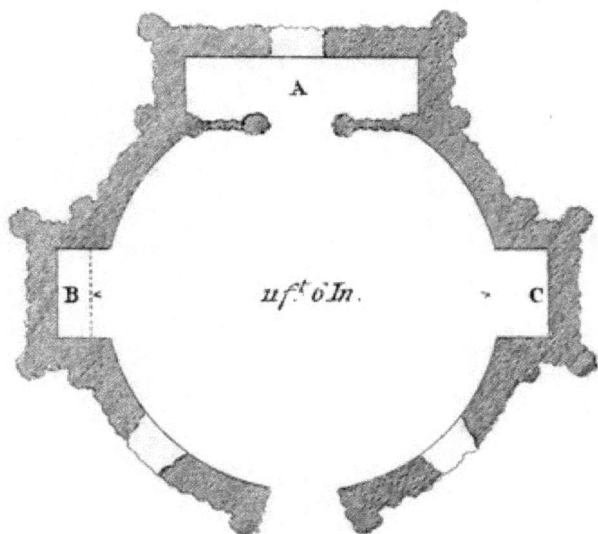

A

B ⟨ ⟩ C

11ft. 6In.

Pl. 7.

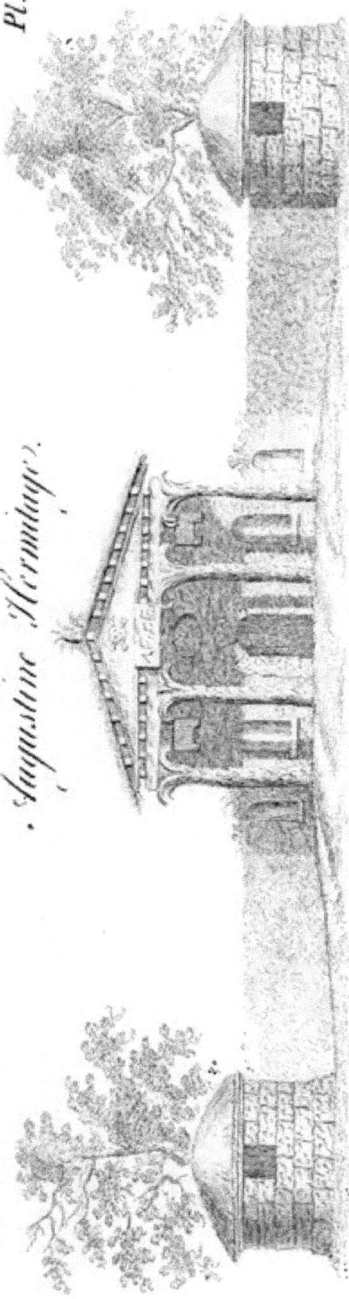

Augustine Hermitage.

Library.

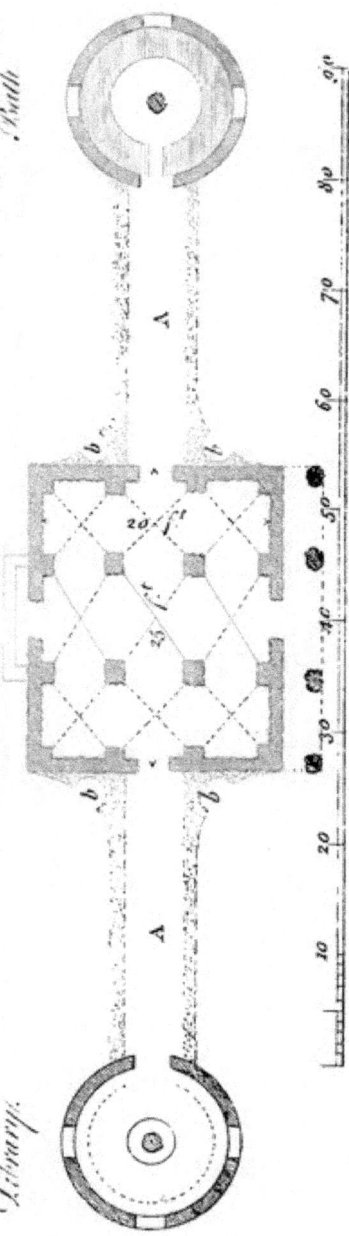

Bath.

Rural Hermitage.

Pl. 3.

Section.   *Elevation.*

5   10   20 f.t

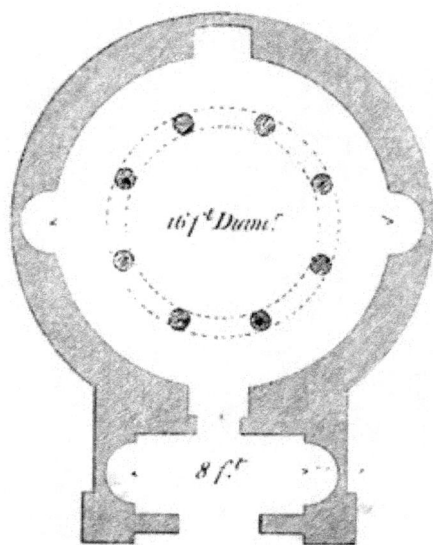

16 f.t Diam.r

8 f.t

Pl. 9.

Modern Grotto.

Elevation.

Section.

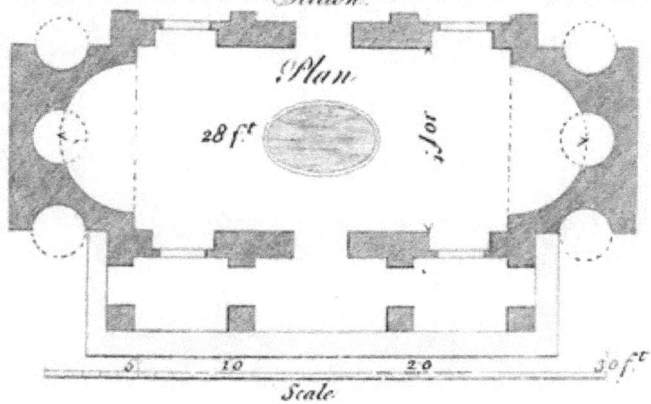

Plan

28 f.^t

for

5   10   20   30 f.^t

Scale.

Gothic Grotto.

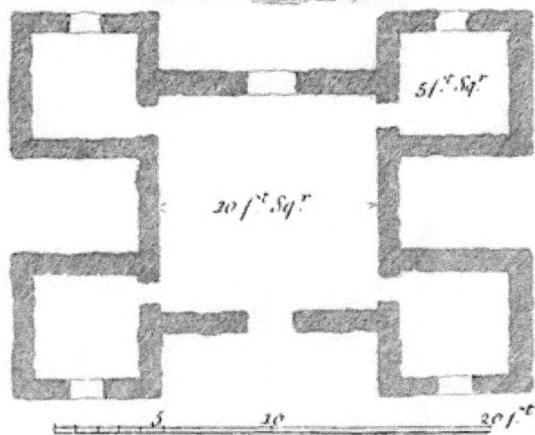

51.ᵗ Sqᵗ

20 f.ᵗ Sqᵗ

5 10 20 f.ᵗ

Pl. 2.

Chinese Grotto.

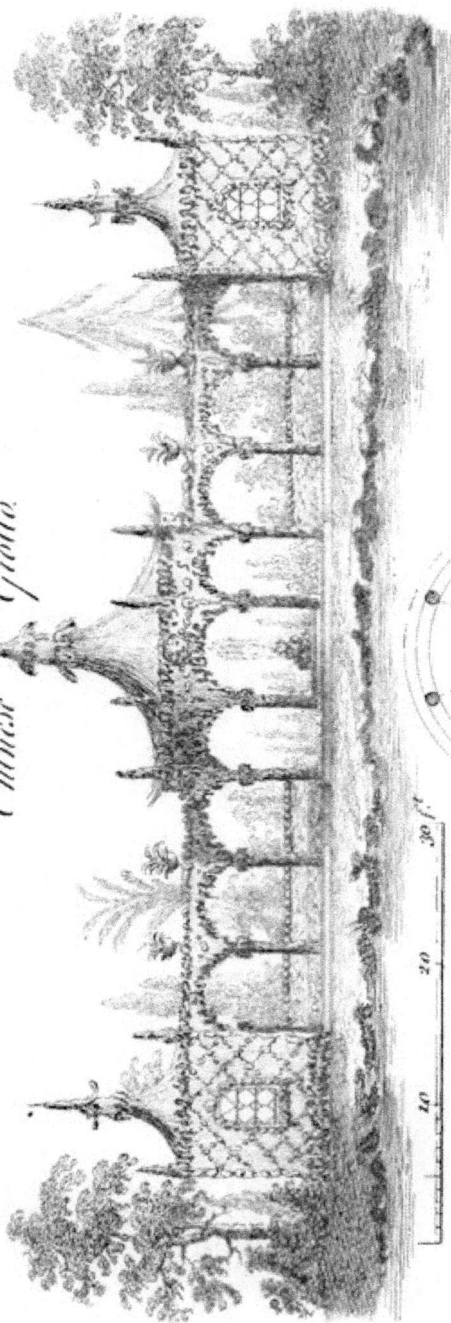

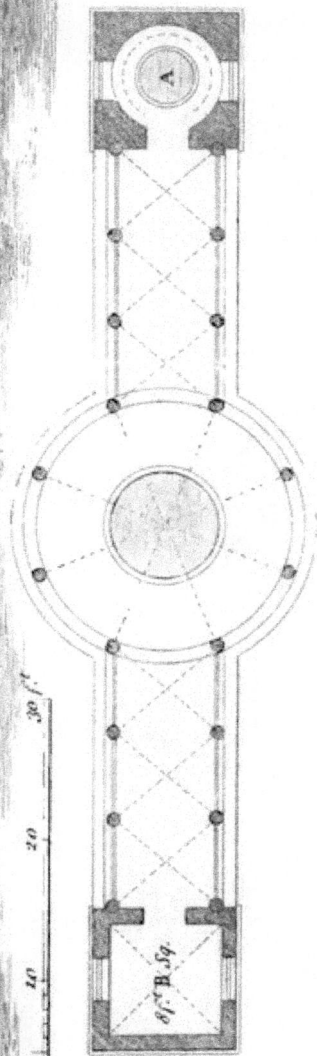

Pl. 12.

Plan to Plate 13.

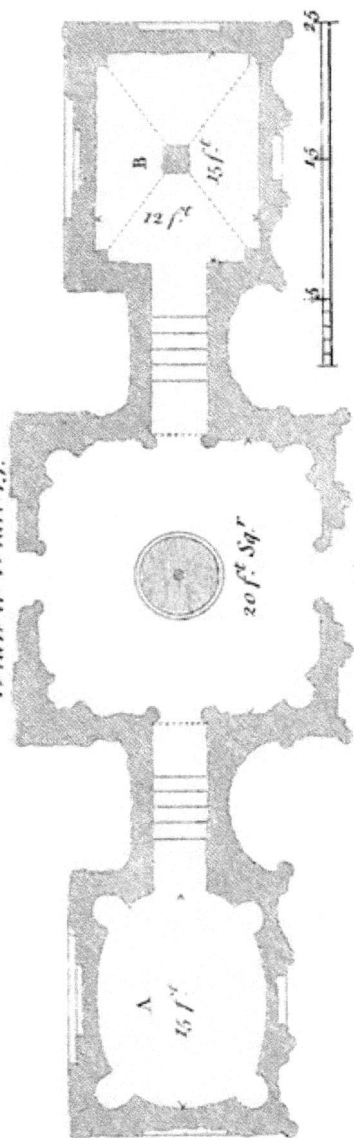

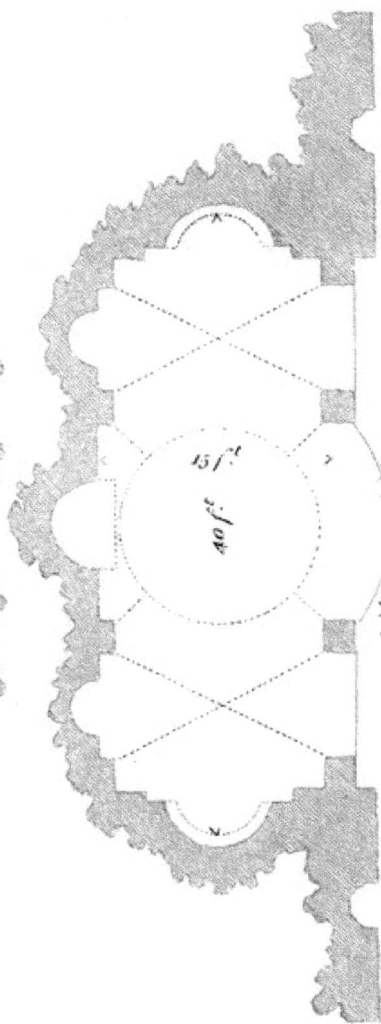

Plan to Plate 14.

A
15 ft

20 ft Sqr

B
15 ft

12 ft

15 ft

40 ft

15 ft

25   15   5

Pl. 13.

Gothic Grotto, with Cascades & Wings, Attached.

Pl. 14.

Rural Grotto

Pl. 15.

A Cascade
with
Fountains

Triumphal Cascade.   Pl. 16.

Pl. 5.

Grotte, Canal, à Gavrudel.

Pl. 18.

Romantic Arches, with Cascades

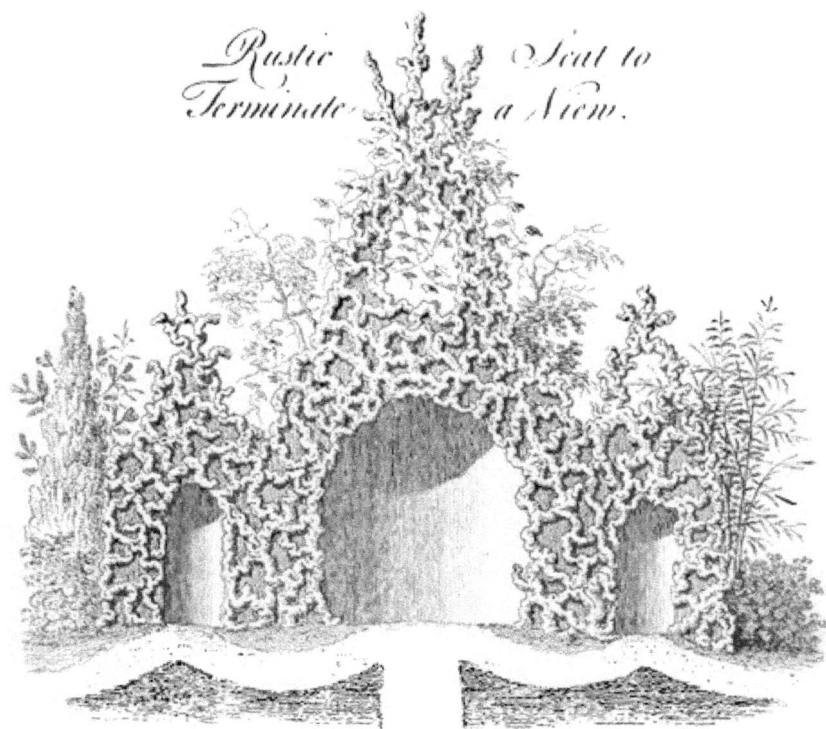

Rustic Seat to Terminate a View.

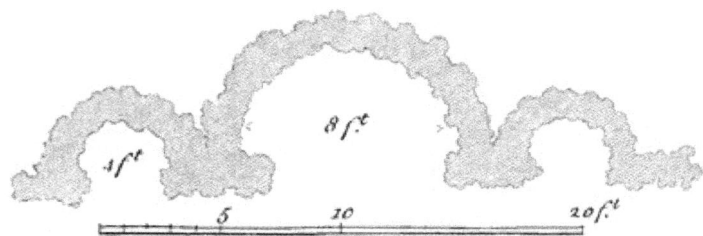

1 f.t   8 f.t   20 f.t

5   10

Grotesque, or Rural Bath.

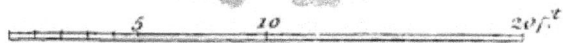

5   10   20 f.^t

Rural Mosque
with
Minarets

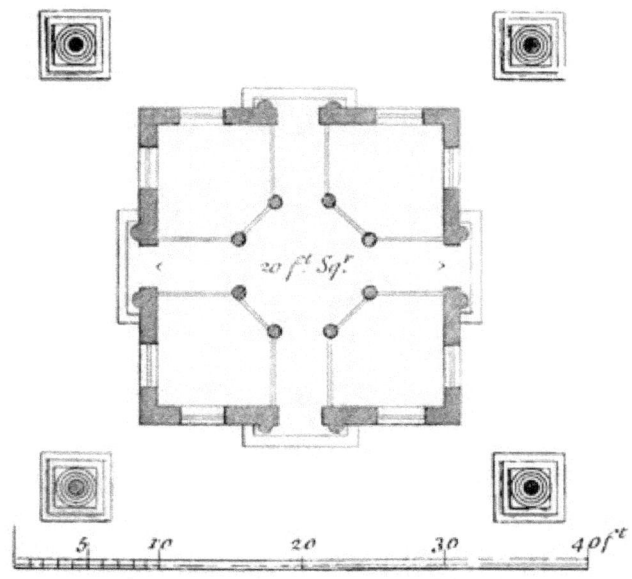

A.   B.

20 f.t Sq.r

5   10   20   30   40 f.t

Pl. 22.

Circular Mosque with Cabinets Attached.

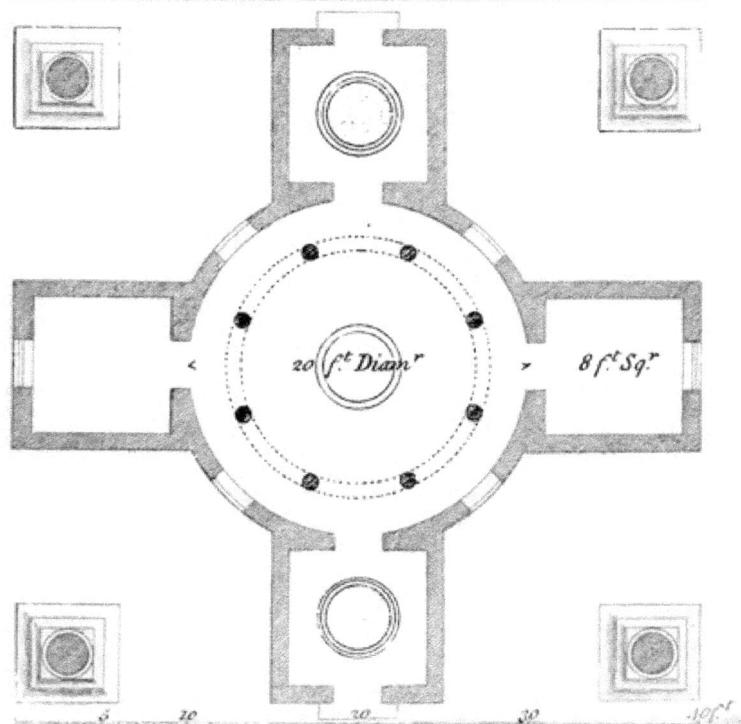

20 f.t Diam.r   8 f.t Sq.r

Pl. 23.

Turkish Mosque with Minarets Attached

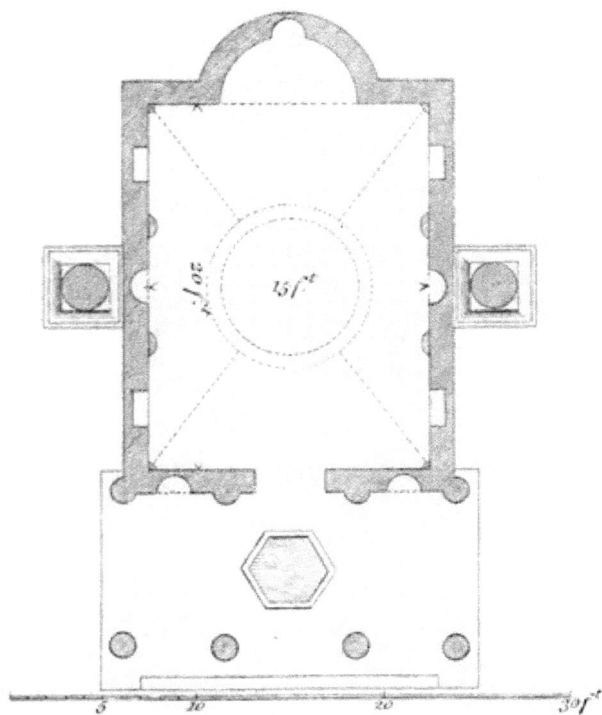

Section to Plate 22.

Section to Plate 23.

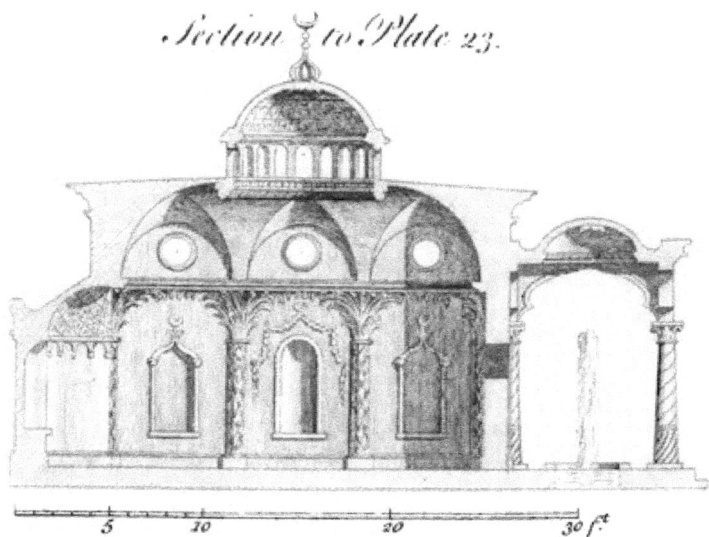

Plan to Plate 26.

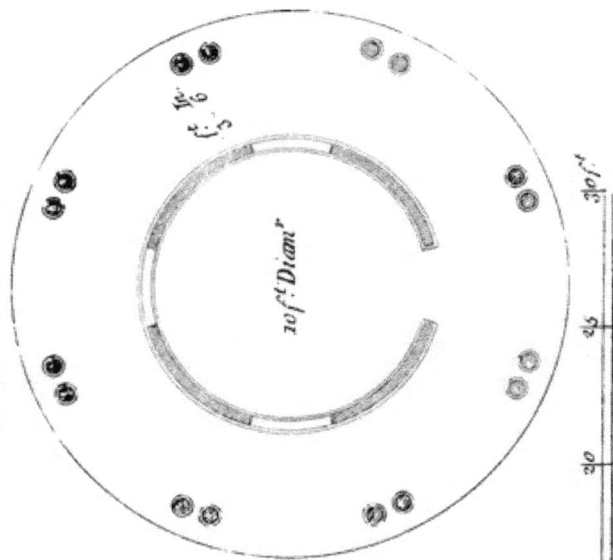

10 ft. Diamr

Plan to Plate 27.

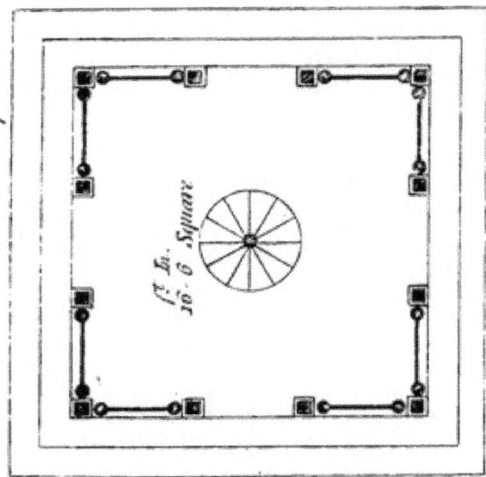

1ft In.
16. 6. Square

Moresque Temple

Moresque Pavillion.

N.º 7

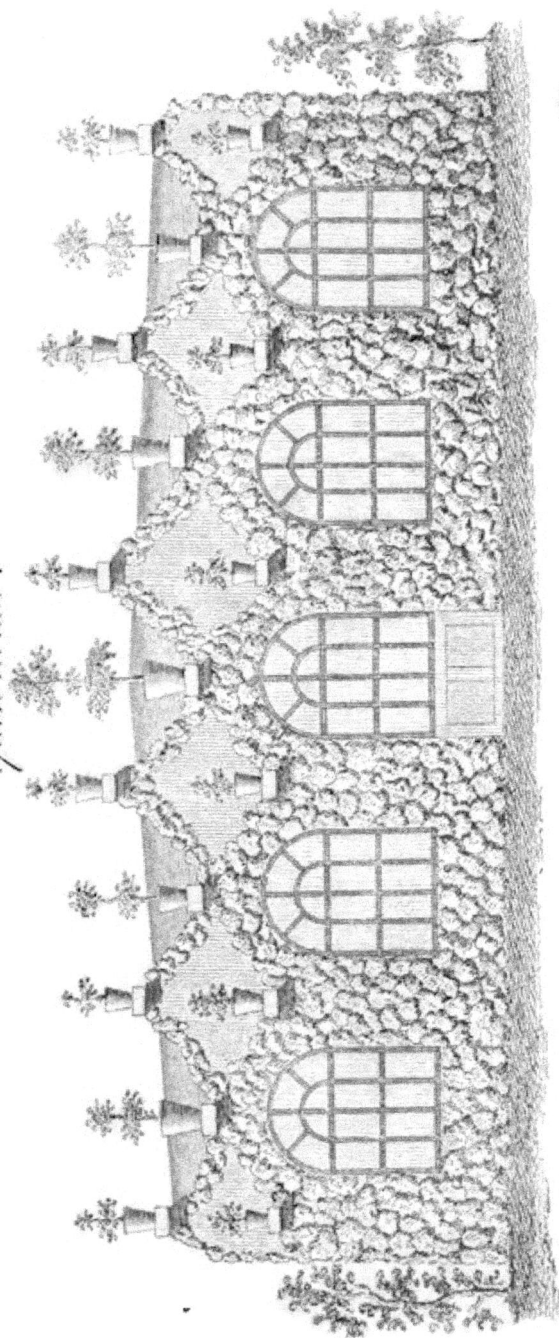

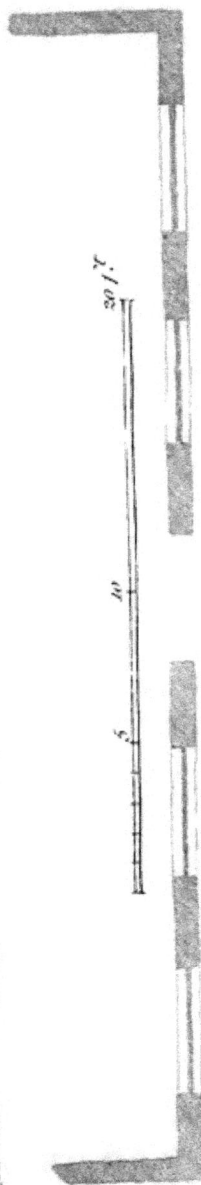

Pl. 22.

Green-house?